LEON COUNTY PUBLIC LIBRARY

WITHDRAWN

3 1260 00747 0316

W9-BGW-478

## DATE DUE

| | | | |
|---|---|---|---|
| | MAY 10.1997 | | |
| JUN.06.1995 | | | |
| | FEB 12.1996 | | |
| AUG.04.1995 | | | |
| OCT.11.1995 | MAR 19.1999 | | |
| | MAY 06.1999 | | |
| NOV.01.1995 | MAY 25.1999 | | |
| FEB.03.1996 | | | |
| FEB.28.1996 | | | |
| APR.09.1995 | | | |
| APR.30.1995 | | | |
| JUL.13.1995 | | | |
| MAR 13.1997 | | | |
| DEC.08.1998 | | | |
| MAY 05.1999 | | | |

Demco, Inc. 38-293

# COPING WITH . . .
# PAPER
# TRASH

# COPING WITH . . .
# PAPER TRASH

Jamie Daniel • Veronica Bonar
Illustrated by Tony Kenyon

Gareth Stevens Publishing
**MILWAUKEE**

**For a free color catalog describing Gareth Stevens' list of high-quality books, call 1-800-341-3569 (USA) or 1-800-461-9120 (Canada).**

**Library of Congress Cataloging-in-Publication Data**

Daniel, Jamie.
    Coping with— paper trash/adapted from Veronica Bonar's Paper rubbish! by Jamie Daniel; illustrated by Tony Kenyon. — North American ed.
        p. cm — (Trash busters)
    Includes bibliographical references and index.
    ISBN 0-8368-1059-7
    1. Waste paper—Juvenile literature. 2. Paperboard—Juvenile literature. 3. Refuse and refuse disposal--Juvenile literature. [1. Paper--recycling. 2. Refuse and refuse disposal. 3. Recycling (Waste)] I. Kenyon, Tony, ill. II. Bonar, Veronica. Paper rubbish! III. Title. IV. Series: Daniel, Jamie. Trash busters.
    TD805.D36  1994                                    93-37688

This North American edition first published in 1994 by
**Gareth Stevens Publishing**
1555 North RiverCenter Drive, Suite 201
Milwaukee, WI  53212,  USA

This edition © 1994 by Zoë Books Limited. First produced as *PAPER RUBBISH!*, © 1992 by Zoë Books Limited, original text © 1992 by Veronica Bonar.  Additional end matter © 1994 by Gareth Stevens, Inc. Published in the USA by arrangement with Zoë Books Limited, Winchester, England.  Published in Canada by arrangement with Heinemann Educational Books Ltd., Oxford, England.

All rights reserved.  No part of this book may be reproduced or used in any form or by any means without permission in writing from Gareth Stevens, Inc.

Series editor: Patricia Lantier-Sampon
Cover design: Karen Knutson

**Picture Credits:**
Sally and Richard Greenhill p. 25; Robert Harding Picture Library pp. 7, 11; Science Photo Library p. 18 (Simon Fraser), p. 22 (Harvey Pincis); Zefa pp. 8, 13, 15, 16, 21, 27.

Printed in the USA

1 2 3 4 5 6 7 8 9 99 98 97 96 95 94

At this time, Gareth Stevens, Inc., does not use 100 percent recycled paper, although the paper used in our books does contain about 30 percent recycled fiber.  This decision was made after a careful study of current recycling procedures revealed their dubious environmental benefits.  We will continue to explore recycling options.

363.72 Dan
0747 0316    NEB
Daniel, Jamie.
4/6/95
Coping with-- paper trash

                    EBS

LEON COUNTY PUBLIC LIBRARY
TALLAHASSEE. FLORIDA

# TABLE OF CONTENTS

Words that appear in the glossary are printed in **boldface** type the first time they occur in the text.

# WHY PAPER IS USEFUL

We use paper every day. We read books printed on paper and use paper to draw pictures. We often pay for items we buy with paper money. We use **cardboard**, a thick, strong paper, to pack things in. We print our computer work on paper.

We use paper for tissues, toilet paper, and paper napkins. At birthday parties, we eat on paper plates. Our birthday presents are wrapped in pretty paper.

Much of what we eat comes packaged in paper or cardboard, and we carry it home from the store in paper bags.

⬥ Paper cups, plates, and napkins are often used for parties.

# PAPER TRASH

More than half of what we throw into the garbage every day is made of paper. This trash is usually gathered and transported by garbage collectors to a **landfill**, where it is **compacted** by heavy machines and then buried. But even after it has been compacted, paper trash takes up a lot of space.

◄ Much of the trash at this landfill is paper or cardboard.

Many people think it is all right to throw paper away since it is **biodegradable**. This means that air and bacteria will make it **decompose**. But because paper at landfills is compacted, air cannot get to it, and it does not decay at all. People have dug up old newspapers that still look new from landfills.

# PAPER LITTER

People often carelessly throw paper candy wrappers or paper drinking cups on the ground or in lakes and streams. This is called **littering**. Paper litter can pile up all over, making parks and streets look dirty and spoiling the **environment** for all of us.

Paper can also block street drains and clog **sewers**. When this happens, rainwater cannot drain away after a storm. And since paper litter gets slippery when it is wet, people and animals can slip on wet paper left on paths and pavement.

⬆ Paper litter is hard to clean up once it has been dropped into a lake or river.

# WHAT IS PAPER MADE OF?

If we tear a piece of paper in half and look at the torn edges, we can see small, hairlike **fibers**. These fibers come from trees and other plants, and they are good for making paper. There are millions of these fibers in every piece of paper.

In addition, cotton and linen **rags** can be **recycled** into paper. The rag fibers are separated and mixed with water to make a **pulp** that looks like oatmeal. The pulp is spread thinly to dry, and the fibers rejoin to make paper. Paper was first made this way in China two thousand years ago.

🔺 Money throughout the world is made from strong rag paper.

# PAPER AND TREES

Most paper is made from **softwood** trees, such as fir and spruce trees. The trees are cut down, and their **bark** is removed. Then the trees are sent to a pulp mill, where they are ground up underwater into a mushy pulp.

**Bleach** and other chemicals are added to the pulp to make it soft and white. Then the pulp is strained, dried, and cut into thick, rough sheets.

These rough sheets of pulp are sent to paper mills, where they are made into a finer and thinner pulp. **Dyes** can be added to make the pulp a new color. The wet pulp is put into a machine that squeezes it onto a wire **mesh**. The mesh is pulled through rollers, and the thin paper that comes out is then dried and rolled up.

⬆ A paper-making machine winds paper onto rolls.

15

# WATER AND WASTE

Making paper uses a lot of water and **energy**. Paper mills use about 80,000 gallons (300,000 liters) of water to make one ton of paper. Most of this water is cleaned and reused at the mill for different batches of paper. When paper is recycled, much less water and energy are needed to make new paper.

➤ Dirty water at a paper plant is left to settle so it can be cleaned and used again.

Unfortunately, when paper is bleached at the mills, some chemicals are released into the air and water as **pollution**. Sometimes these chemicals are carried out of the mill in water and paper fibers, and then the chemicals pollute rivers and lakes elsewhere.

# SAVING TREES

If we recycle more paper and do not use as much to begin with, fewer trees will need to be cut down. It takes about seventeen trees to make just one ton of paper.

Many softwood trees are planted on purpose so they can be used for making paper. We are presently cutting down so many trees, however, that we still cut down more than we plant.

▲ This machine cuts down trees and then strips off their branches and bark.

Sometimes people grow softwood trees on land that was once covered by other types of trees, called **deciduous** trees. Many animals and plants thrive in deciduous forests. When cone-bearing softwood trees replace deciduous trees, the environment for these life-forms changes. Softwood trees grow thick and close together, so less sunlight filters through the treetops. Many of the plants and animals that once lived in the area no longer have homes and cannot survive in the new environment.

19

# SORTING PAPER FOR RECYCLING

If we recycle paper instead of simply throwing it away, we will help save trees and the rest of our environment. But not all paper can be recycled. Paper packaging with bits of food on it and any paper coated with foil, plastic, or wax are not recyclable.

These **bales** of paper are on their way to a pulp mill for recycling.

Other types of paper, however, can be recycled easily. The paper must first be sorted into separate types and then tied into bundles. In some communities, recyclable paper is picked up with other trash. In other places, people take the paper to a **recycling center** themselves. The people who work at the center send the paper off to a pulp mill.

# RECYCLING PAPER

Different papers are different in quality. Drawing and writing paper are usually high-grade papers, while newspapers are printed on low-grade paper. Low-grade paper is recycled into more low-grade paper; high-grade paper is recycled into more high-grade paper.

◆ Waste paper and cardboard are being made into pulp for new low-grade paper.

When waste paper arrives at the pulp mill, any ink or bits of metal on it are removed. Water is added, and the paper is mashed into a pulp. A little pulp from new wood is added for better quality. Then the pulp is dried, packed, and sent to a paper mill to make new paper.

The next time you need to buy paper, look for paper that has been recycled. You'll be doing the trees — and yourself — a big favor!

# REUSING PAPER

One way to avoid wasting paper is to reuse paper products. Envelopes, paper bags, and wrapping paper, for example, can be used more than once. Notes and shopping lists can be written on the blank backs of printed sheets of paper that are no longer needed. Old newspapers can be used to protect young plants in the garden and keep them moist.

Rolled up newspapers can be used to make soft beds for animals or paper logs for a fireplace. Old newspaper can also be used to make **papier mâché**, a thick paste of glue, water, and paper. The paste can be sculpted into gifts such as puppets, masks, and bowls for parents and friends.

🔺 These brightly painted masks are made of papier mâché.

# USING LESS PAPER

The best way to avoid wasting paper is also the easiest — just don't use so much! Use a cloth towel to clean up spills instead of a paper towel. Use washable plates and cups instead of paper products at your next party or picnic. And try to avoid buying products with unnecessary packaging.

Take a cloth, plastic, or paper bag from home to the store instead of using new bags. Print out your work from a computer printer only when the work is complete and correct. It is not difficult to use paper wisely instead of wastefully. And saving paper will help preserve precious trees.

▲ Look carefully at the computer screen when you work, and print the work only when you are sure it is ready.

27

# GLOSSARY

**bales:** large bundles of goods that have been pressed together to take up as little space as possible.

**bark:** the protective outer covering of trees. Tree trunks, branches, and roots are usually covered with bark.

**biodegradable:** able to be broken down by air and bacteria.

**bleach:** a chemical that removes color from many substances. Bleach is often used in the process of making paper.

**cardboard:** a very thick, low-grade paper used to make boxes and other sturdy paper items.

**compact:** (v) to flatten and press together large amounts of trash so it takes up less space in a landfill. Big machines compact trash at landfills.

**deciduous (trees):** trees with leaves that change color and fall off in autumn or winter. Oak, maple, and hickory are examples of deciduous trees.

**decompose:** to decay or rot.

**dye:** a substance used to change the color of things. Dyes can be added to make paper a certain color.

**energy:** the power necessary to do work. We eat food to get the energy we need for our bodies to perform different tasks. Machines get energy from different types of fuel.

**environment:** the surroundings in which people or animals live. By trying to cut down fewer trees for paper and paper products, we help protect both the trees and the rest of our environment.

**fibers:** thin strands of an artificial or natural substance.

**landfill:** a big hole in the ground where trash is dumped and then covered with soil.

**litter:** the trash many people carelessly throw on the ground or in other places, such as lakes and streams.

**mesh:** a net or screen.

**papier mâché:** a paste made of paper mixed with water and glue.

**pollution:** the gas, smoke, trash, and other harmful substances that ruin our environment.

**pulp:** a soft, mushy mixture of wood fiber and other substances used to make paper.

**rag:** a scrap of torn cloth.

**recycle:** to make new products from old products that have already been used. Many metal, glass, plastic, wood, and paper products can be recycled.

**recycling center:** a place where materials are gathered for recycling.

**sewers:** drains or pipes, usually under the ground, that are built to carry away waste materials.

**softwood:** wood taken from trees that have needles and cones, such as fir and spruce trees.

# PLACES TO WRITE

Here are some places you can write for more information about paper and how it can be recycled. Be sure to give your name and address and be clear about the information you would like to know. Include a self-addressed, stamped envelope for a reply.

Paper Stock Institute
c/o The Institute of Scrap
   Recycling Industries
1325 G Street NW
Suite 1000
Washington, D.C.  20005

American Forest & Paper
   Association
1250 Connecticut Ave. NW
Washington, D.C.  20036

Greenpeace Foundation
185 Spadina Avenue
Sixth Floor
Toronto, Ontario
M5T 2C6

# INTERESTING FACTS ABOUT PAPER

**Did you know . . .**

- that paper is sometimes used to make clothing?  For instance, the gown you change into at the doctor's office is made of paper.

- that before the invention of paper, the ancient Egyptians used to write on long strips of papyrus, which is a plant leaf?  Other cultures wrote on cloth.

- that the first printing press was invented by Johann Gutenberg in about 1650?  Before this time, all books had to be printed by hand, so books were rare and expensive.

- that a special symbol is printed on new paper if it has been made from recycled paper?  This way, recycled paper can be easily identified.

# MORE BOOKS TO READ

*Earthwise at Home: A Guide to the Care and Feeding of Your Planet.* Linda Lowery (Carolrhoda)

*Paper Through the Ages.* Sharon Cosner (Carolrhoda)

*The Story of Paper.* Odile Limousin (Young Discovery Library)

*Trash Attack: Garbage and What We Can Do About It.* Candace Savage (Firefly Books)

*Why Does Litter Cause Problems?* Isaac Asimov (Gareth Stevens)

*Wood and Paper.* Jacqueline Dineen (Enslow)

# INDEX